This edition first published in 2005 by
Sea-to-Sea Publications
1980 Lookout Drive
North Mankato
Minnesota 56003



ISBN 1-932889-23-X

Printed in China



Library of Congress Control Number: 2004103718

2 4 6 8 9 7 5 3

Published by arrangement with the Watts Publishing Group Ltd, London

Author: Fiona Campbell

Edited by Matthew Parselle

Designed by Victoria Illiffe

Illustrated by Michael Evans
Photography by Steve Shott
Series editor: Paula Borton
Series designer: Robert Walster

# LOOK AND MAKE

# PAINTING




SEA-TO-SEA

*Mankato Collingwood London*

# Before you begin

Make sure that you have all the things you need. Cover the surface you are working on with lots of old newspaper.

## Things you may need:

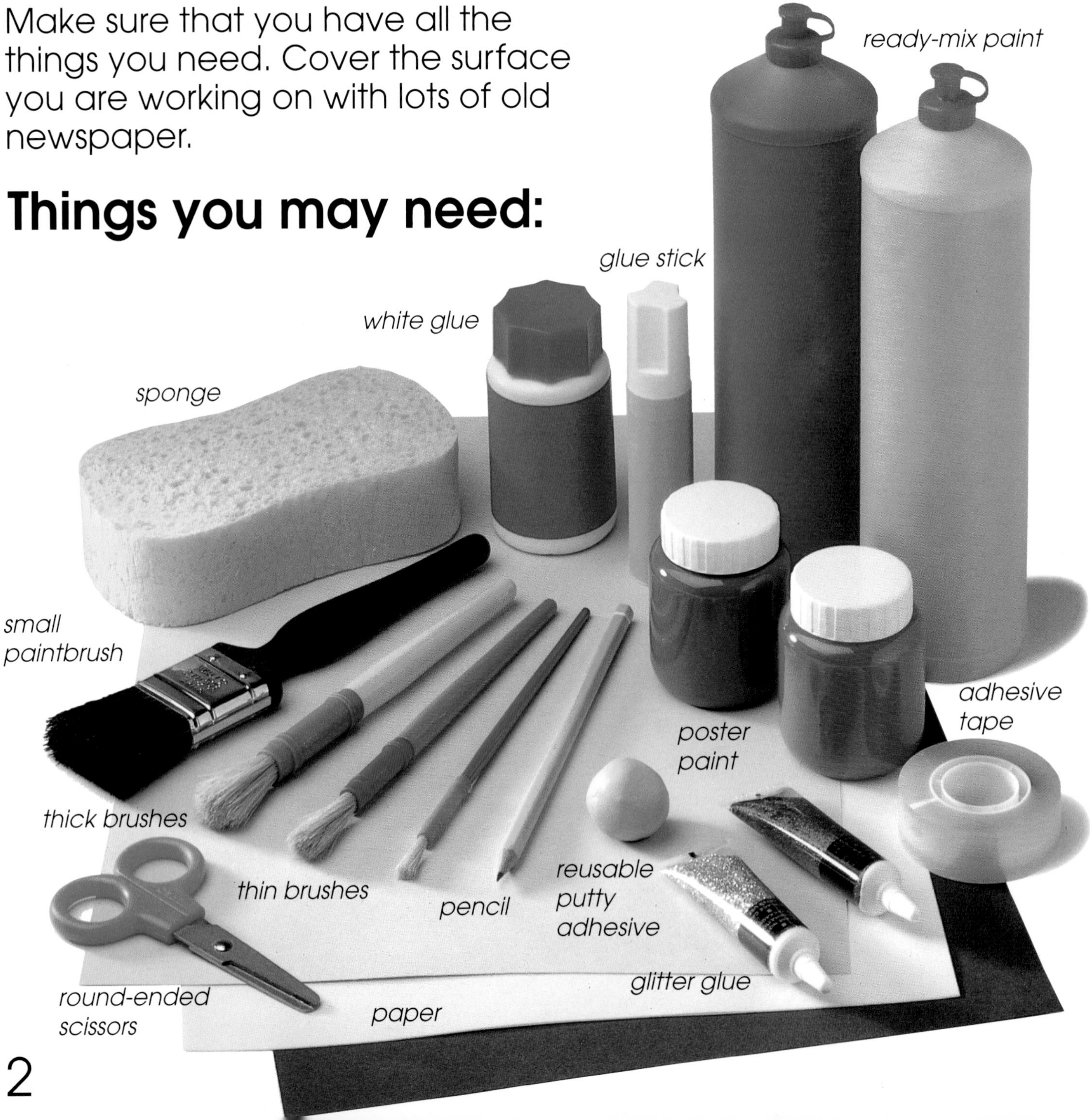

# Be prepared

Half-fill a jar with water. You'll need to fill it with clean water many times while you are painting.

Wash your brush in the jar frequently as you paint. Try to keep your hands clean, too.

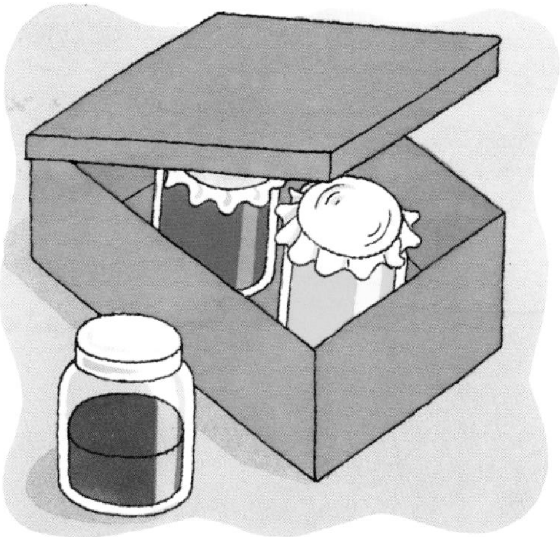

Keep your paints in pots with lids or cover them with plastic wrap and put them in a box with a lid.

# Mixing colors

You can mix two different colors to make another color.

red + green = brown

blue + yellow = green

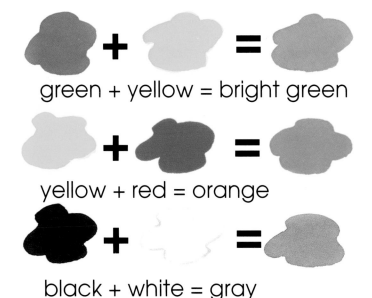

green + yellow = bright green

yellow + red = orange

black + white = gray

# Spoon creature

## You will need:

*bright paper*

*old wooden spoon*

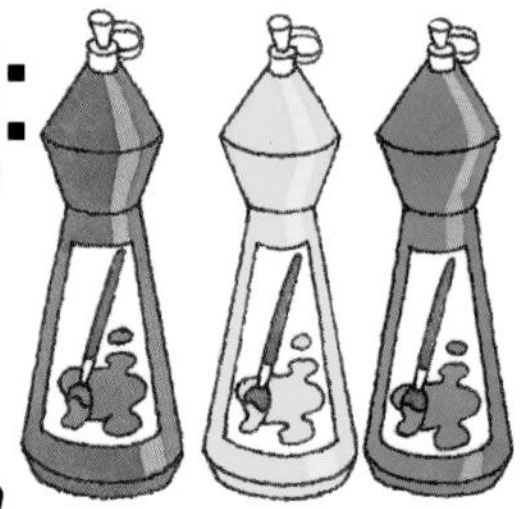

*paints*

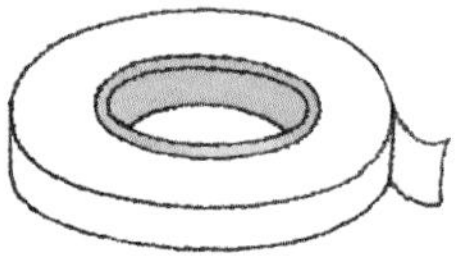

*masking or adhesive tape*

*white glue*

*Add a red tongue to make a snake.*

## 1.

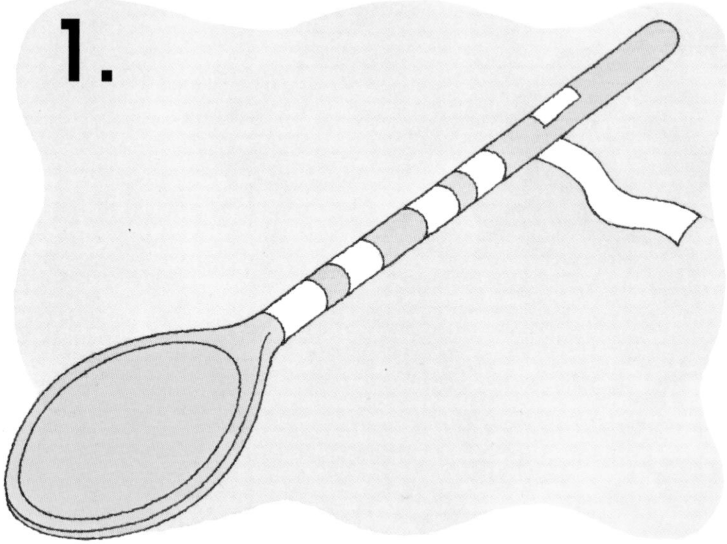

Stick pieces of tape around the spoon. Leave some gaps.

## 2.

Paint all over the spoon. Leave it to dry. Peel off all the tape.

*Glue the wings to the spoon to make a dragonfly.*

**3.**

Paint on big eyes and lots of patterns on the stripes.

**4.**

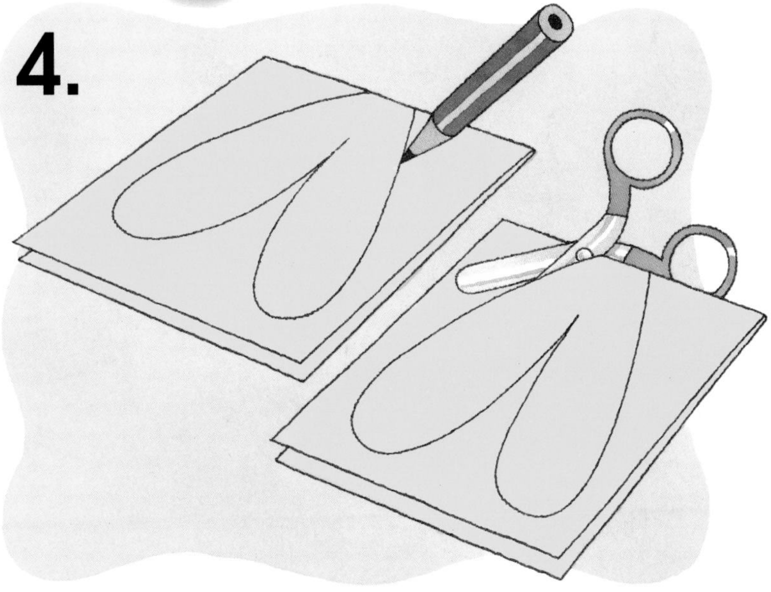

Draw wings on folded paper. Cut through on both sides.

# Rollercoaster

## You will need:

*long piece of paper*

*small potato, cut in half*

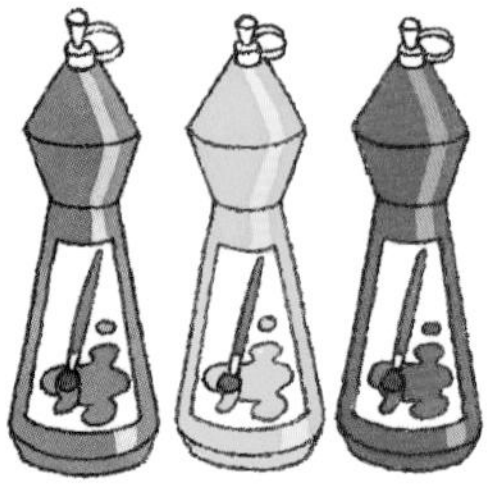

*paints*

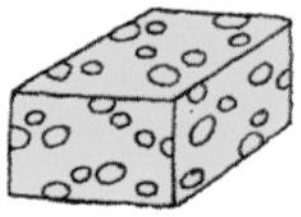

*small blocks of sponge*

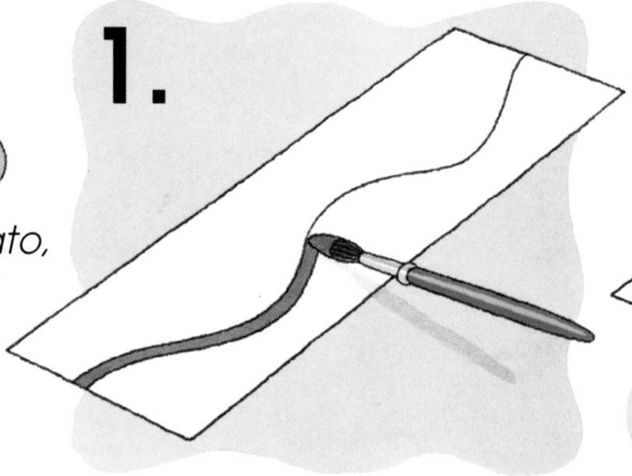

**1.** Draw a wavy line on your paper with a pencil. Then paint over it.

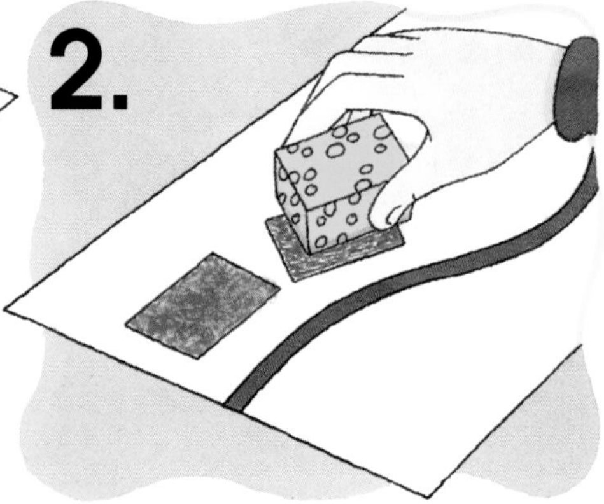

**2.** Dip a sponge into some bright paint and print lots of shapes along the line.

**3.**

Dip one half of the potato in paint and print lots of heads. Add faces to them.

**4.**

Dip the other half of the potato in some paint and print lots of wheels.

**5.**

Paint lots of long, thin lines on your rollercoaster and leave it to dry.

# Spotty dotty turtle

## You will need:

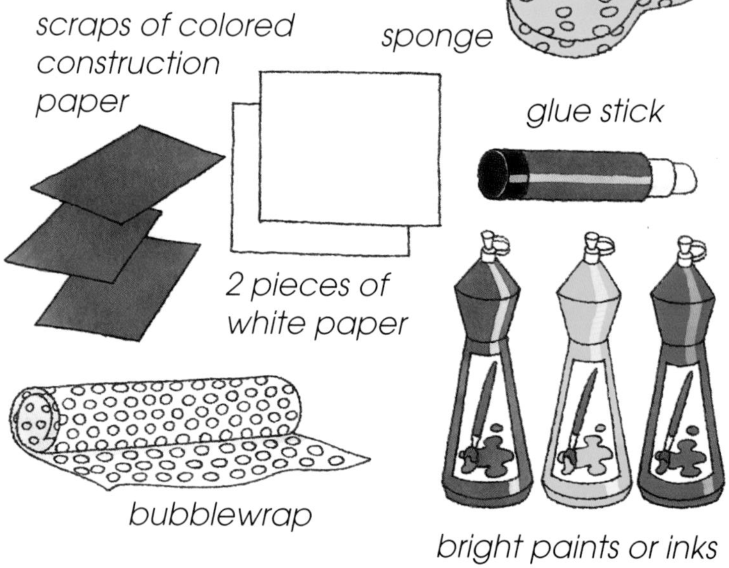


**1.**

Use a wet sponge to make the paper damp all over. Add lots of splodges of paint or ink.

**2.**

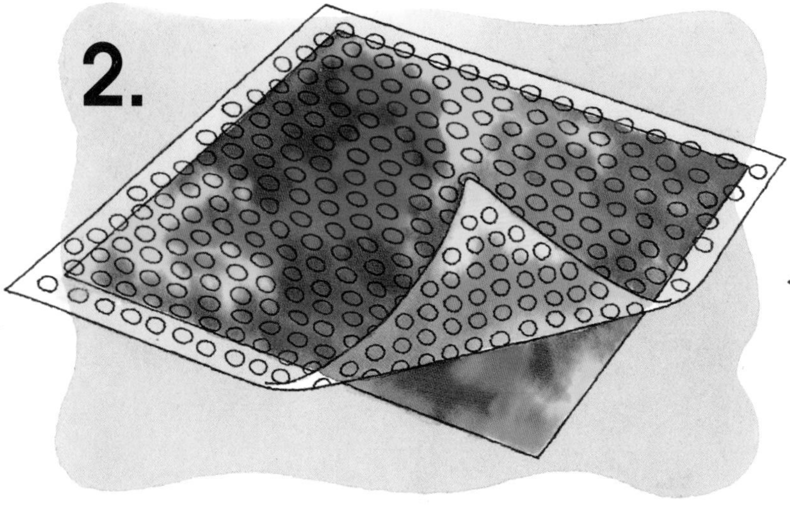

Lay the bubblewrap on top. Carefully rub your hand over it then lift off the bubblewrap.

**3.**

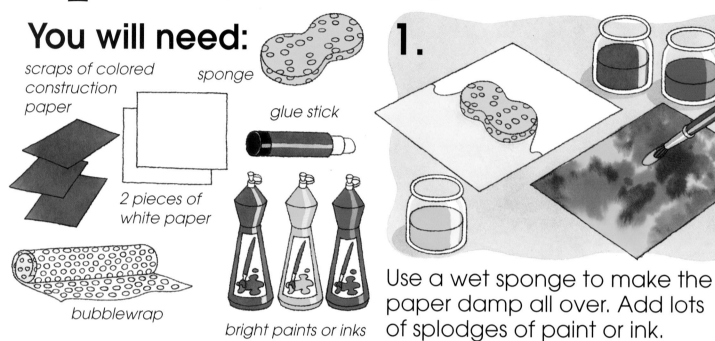

Lay the bubblewrap, paint-side down, on a new piece of paper. Rub over it then lift it off.

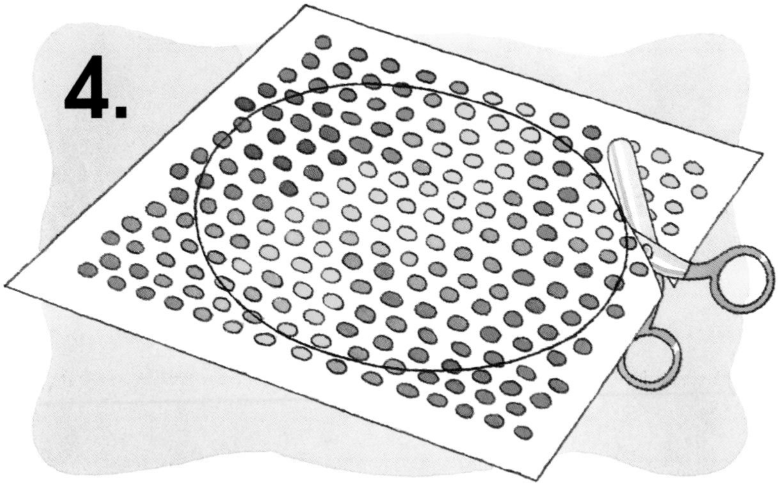

**4.**

**5.**

Leave it to dry. Draw a big oval shape for the turtle's shell. Then cut it out carefully.

Cut out a head, flippers, and the tail from the scraps of paper. Glue them onto the back of the shell.

*Try using plastic wrap instead of bubblewrap.*

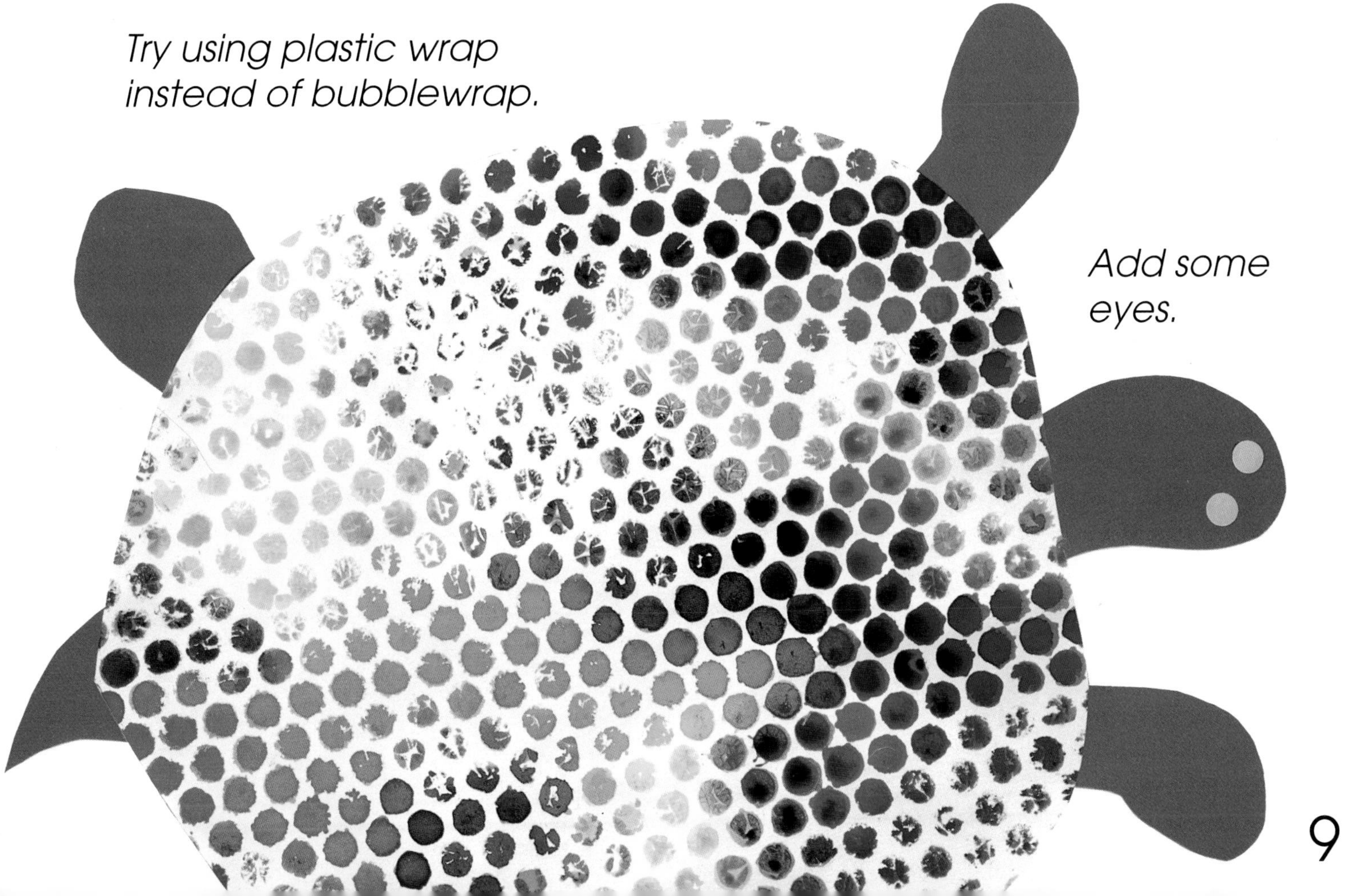

*Add some eyes.*

# Spider's web

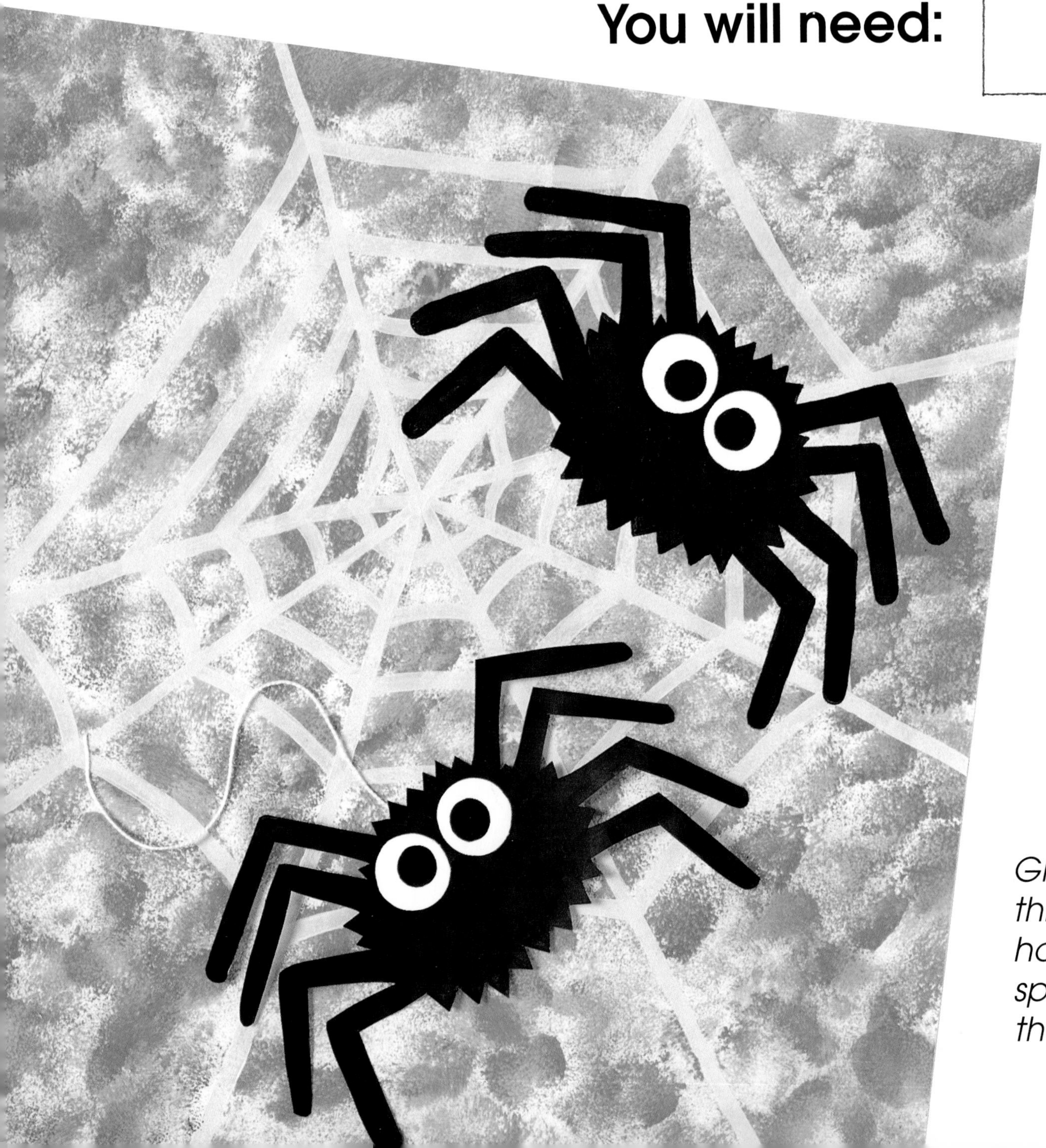

## You will need:

*large piece of paper*

*old plate (for mixing paint)*

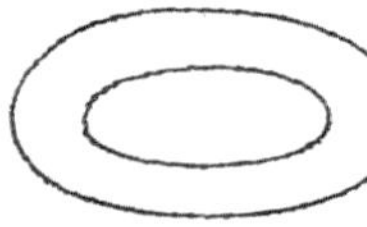

*paints*

*small chunky sponge*

*Glue on some thread and hang another spider from the web.*

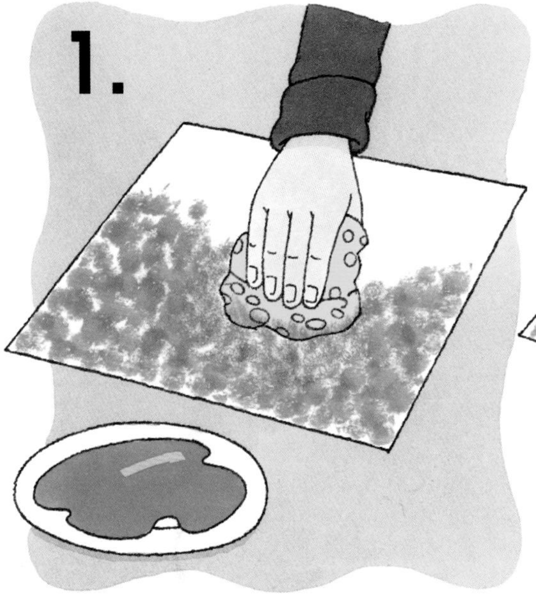


Mix some thin paint and sponge it onto the paper. Leave it to dry.

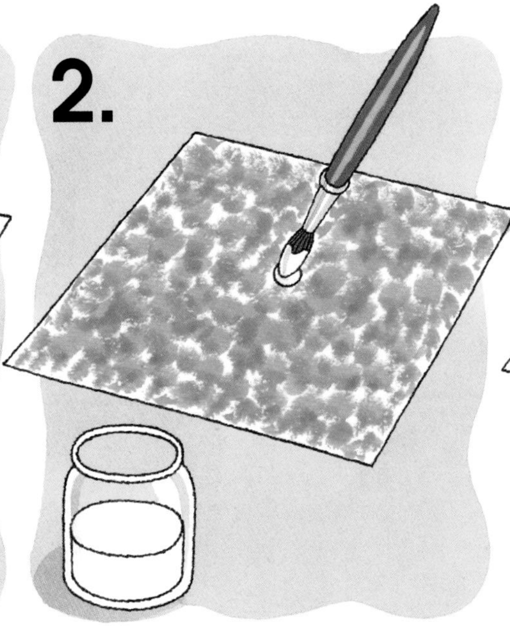


Paint a dot somewhere near the middle of your paper.

Paint some lines out to the edge of your paper with a thin brush.

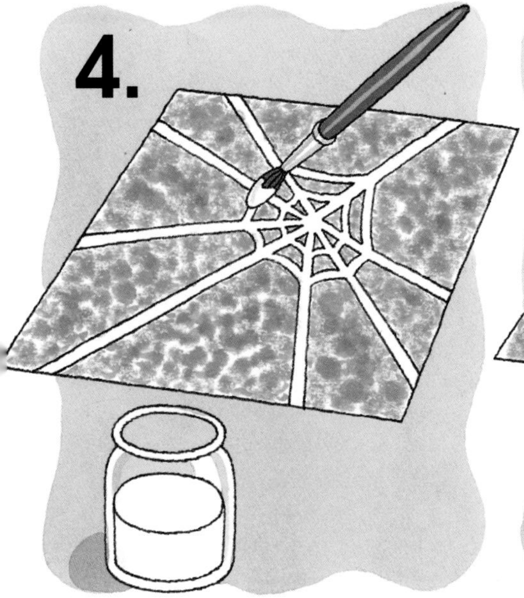


Starting near the middle, add curved lines to make a web.

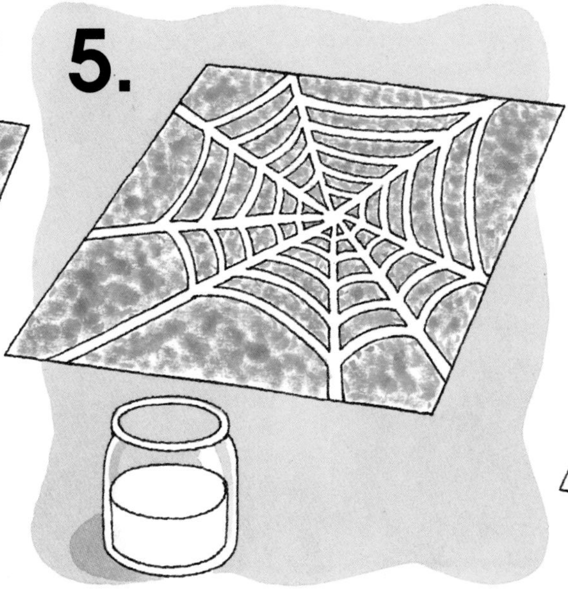


Carry on adding more lines as shown. Leave it to dry.

Paint a big spider somewhere on your web. Add some eyes.

# Into space

## You will need:

black construction paper

reusable putty adhesive

paints

old toothbrush

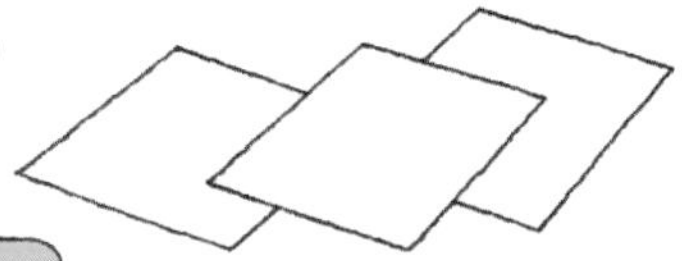

scrap paper

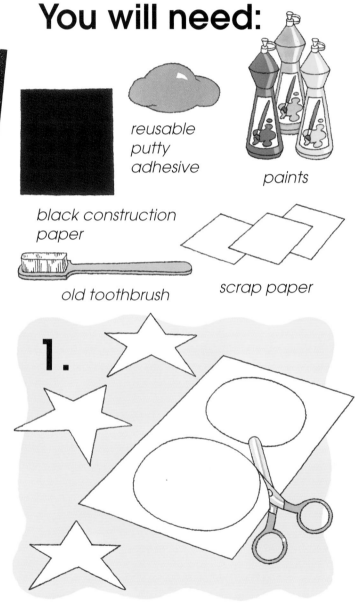

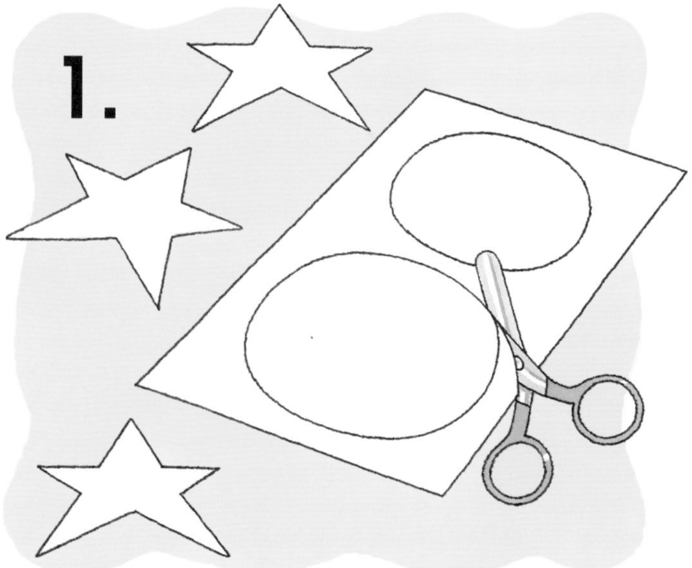

1. Draw some stars and one or two circles for planets on the scrap paper. Cut them out.

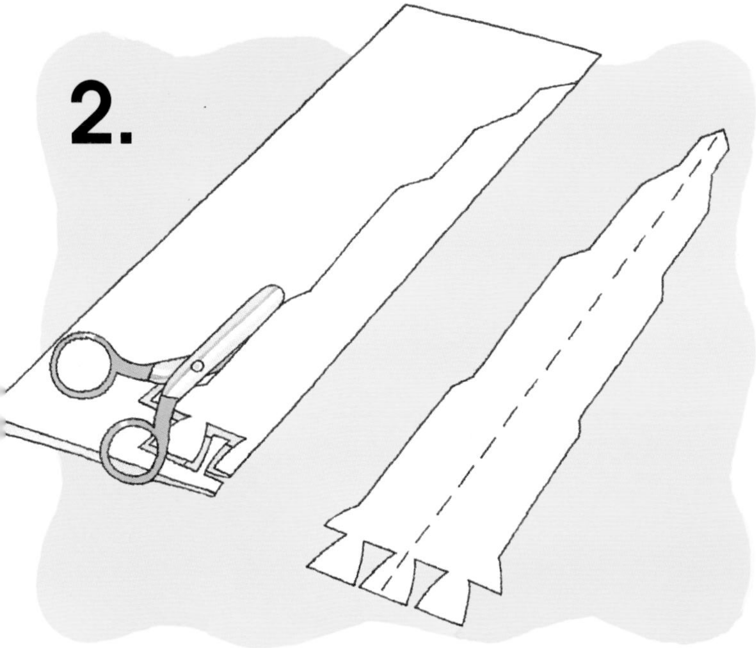

Draw half a rocket on folded paper. Cut through both sides.

Stick all the shapes to the black paper with the putty adhesive.

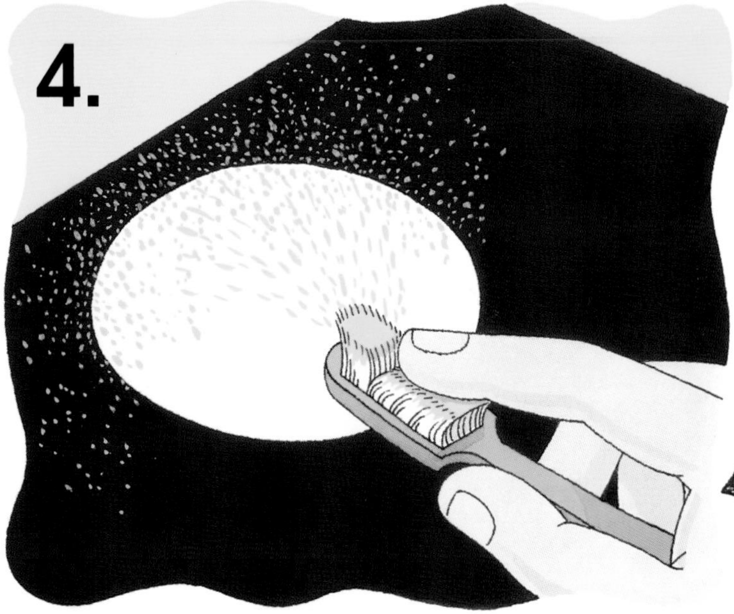

Dip the toothbrush in paint. Drag your finger over the brush toward you to splatter the paper.

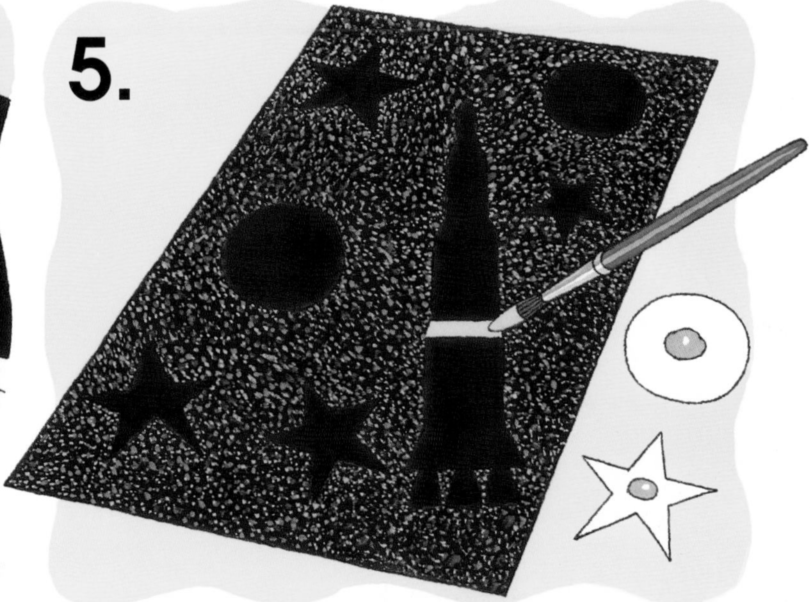

Splatter on more colors. Leave it to dry then pull off the pieces. Paint the shapes using a thin brush.

# Bowl of flowers

## You will need:

large sheet of paper

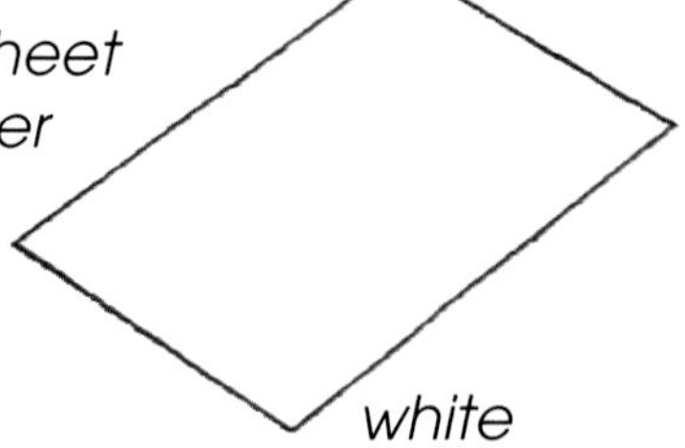

white glue

paper plate

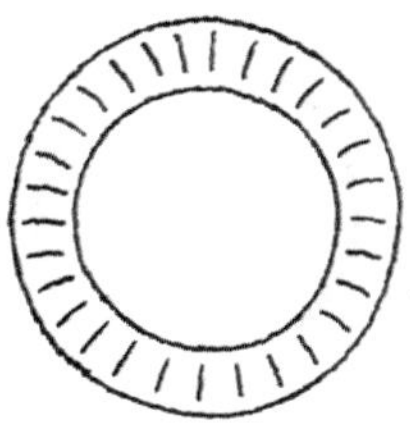

bright paints

different round objects (mugs, cups, yogurt cartons)

*You could add spots and stripes to the flowers.*

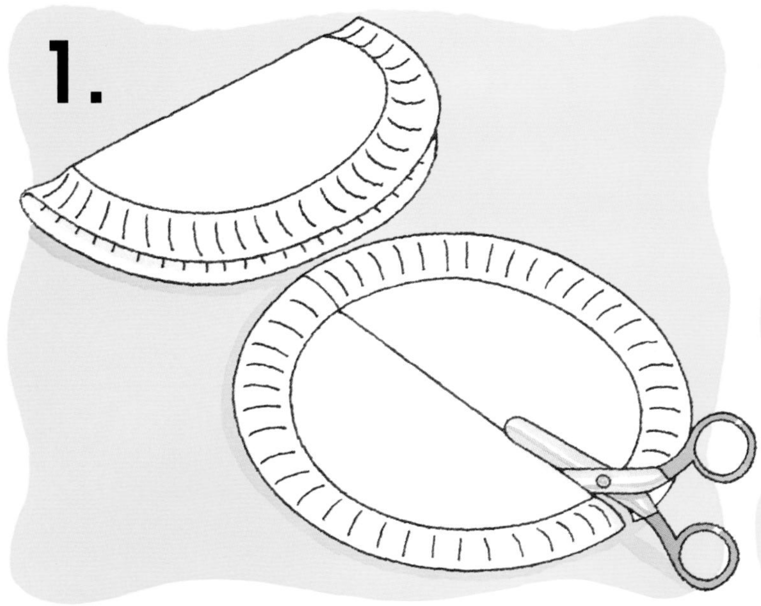

Fold the paper plate in half and open it out again. Use scissors to cut it carefully along the fold.

Paint a bright pattern on one half of the plate. Glue it near the bottom of your paper.

For the middles of the flowers, draw around the round things. Leave lots of space between.

Paint the middles and add petals around them. Add stems and leaves in the spaces.

# Monster meal

## You will need:

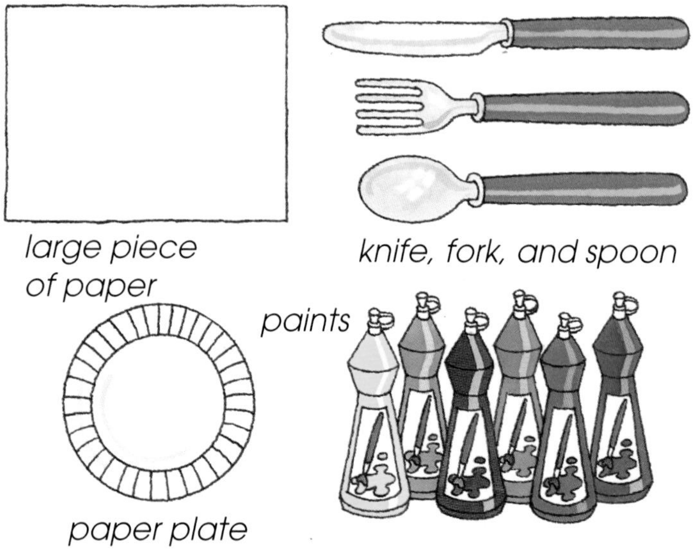


**1.**

Put the knife, fork, and spoon on the paper as shown. Draw around them.

**2.**

Draw lines across and down your paper for a tablecloth.

**3.**

Paint the knife, fork, and spoon and the squares on the tablecloth.

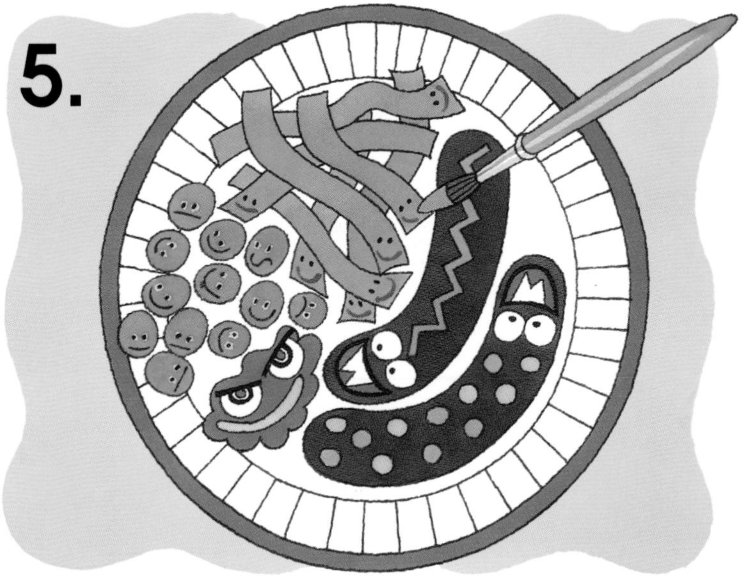


Paint a pattern around the edge of the paper plate. Add a huge, tasty meal.

Add faces to the food to make each thing look like a scary monster.

*You can glue the plate to the tablecloth if you like.*

# Snake in a basket

## You will need:

- large piece of paper
- white glue
- small piece of paper (about the size of this book)
- flour
- small strip of thin cardboard
- paints
- tablespoon
- old teaspoon
- water

1.

Use scissors to cut some V shapes in one end of the small strip of cardboard.

2.

Mix a tablespoon of flour with a little water. Add a teaspoon of glue and mix in some paint.

3.

Use a big brush to paint the small piece of paper with the mixture. Do this very quickly.

4.

Drag the cut end of the cardboard across the paint to make patterns. Leave to dry.

5.

Place this book at the bottom of the big paper. Draw a snake above it. Then paint the snake.

6.

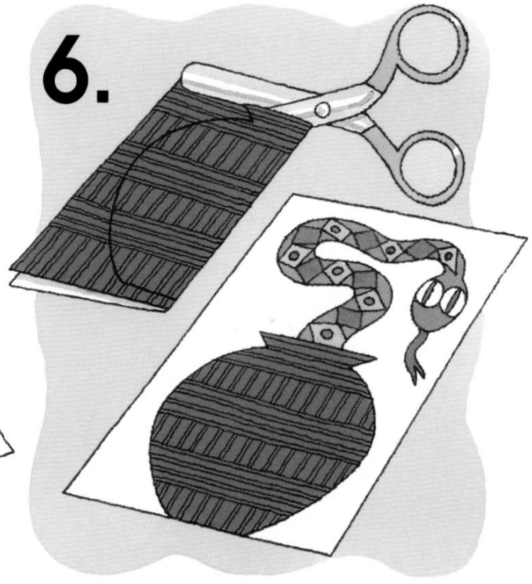

Fold the small paper. Draw half a vase and cut through on both sides. Glue it down.

# Spooky sunset

## You will need:

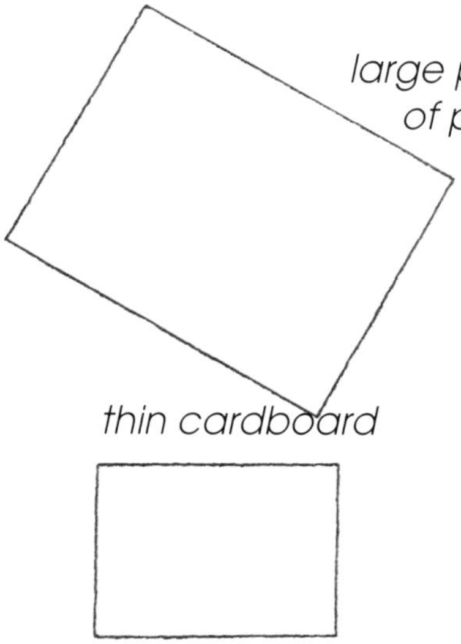

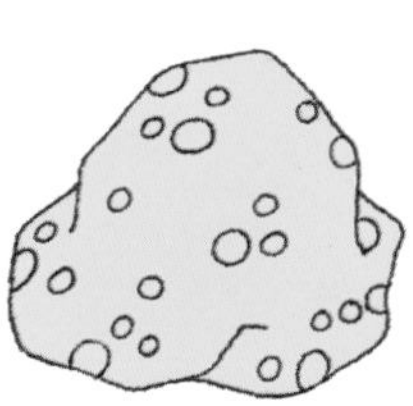

**1.**

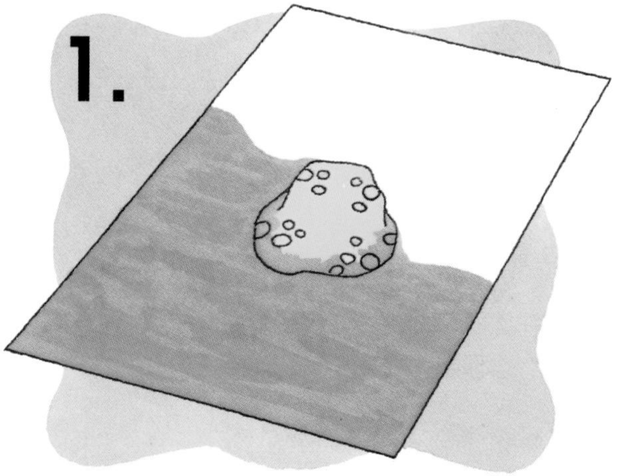

Mix some yellowy-orange paint. Dip the sponge in it and rub it across the large paper. Let it dry.

**2.**

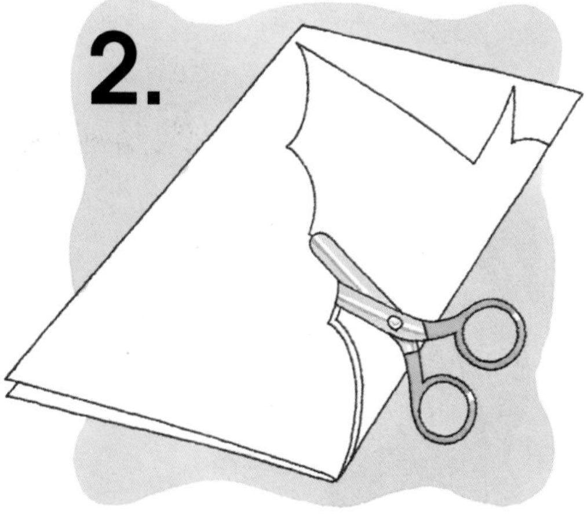

Fold the card. Draw half a bat along the fold. Cut it out through both sides.

**3.**

Open out the bat and lay it on the painted paper. Draw around it.

**4.**

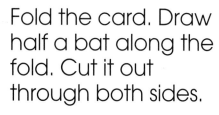

Draw another bat shape on the paper. Then paint both bats black. Add scary eyes and teeth.

5.

Mix some very watery black paint. Drip blobs of it at the bottom of your picture.

6.

Put one end of the straw on the blobs and blow hard to make the spooky trees.

*You could make the second bat smaller.*

# Flashy cat

## You will need:

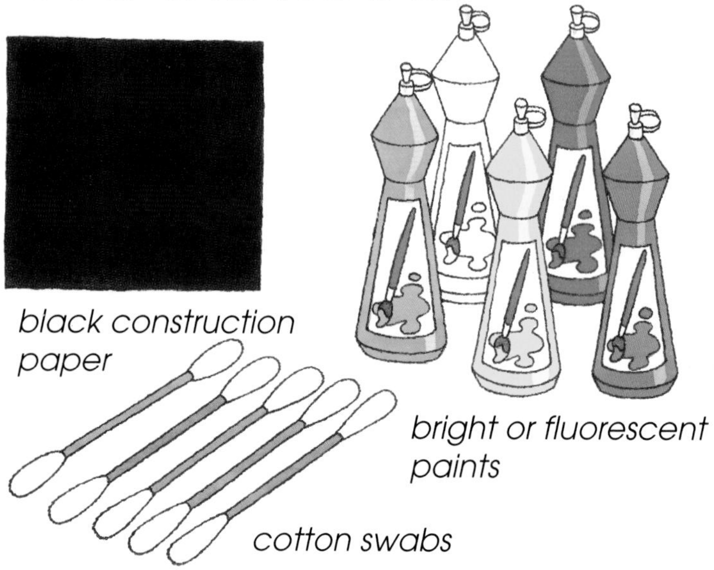


1.

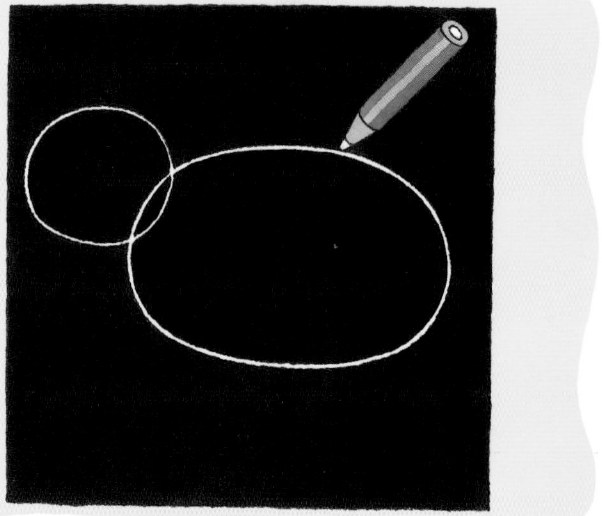

Draw a circle for the head and an oval for the body of the cat.

2.

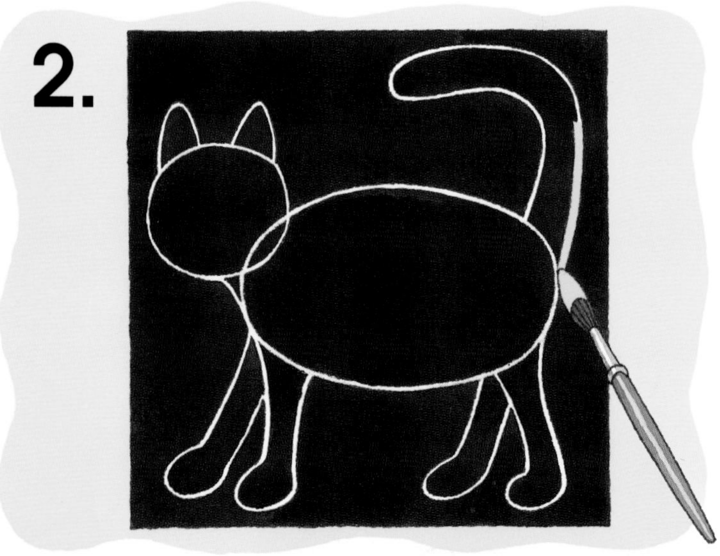

Add some ears, legs, and a tail. Paint the outline of the cat.

3.

Paint some eyes, a nose, and some stripes on its body.

Paint between some of the stripes and add lots of circles.

Dip cotton swabs into some paint and use them to print lots of dots.

*Try painting other animals and decorating them with different patterns.*

# Flashy fish

## You will need:

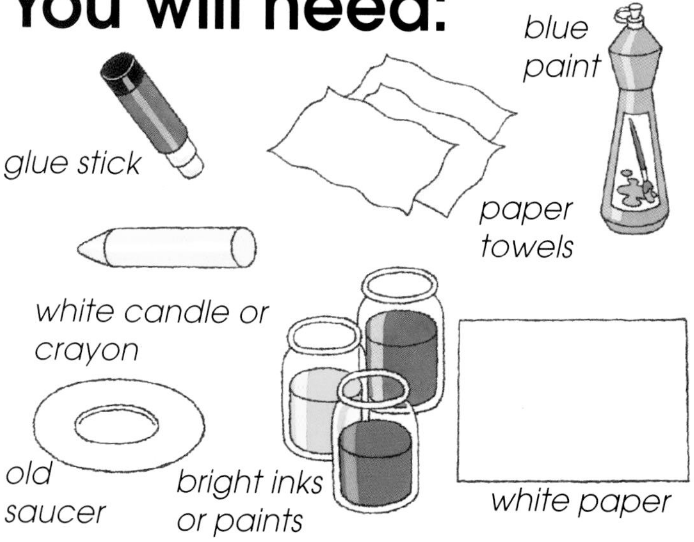

Dampen two paper towels. Drip watery paint or ink onto them. Leave them to dry.

While the towels are drying, draw wavy lines across your paper with the candle.

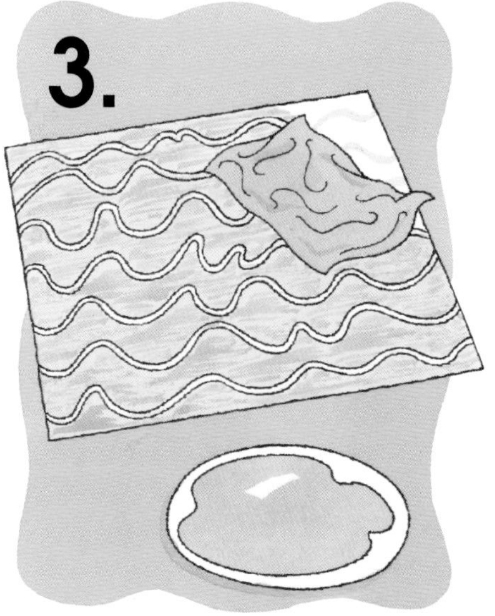

Mix blue paint with lots of water. Wipe it across your paper with a paper towel.

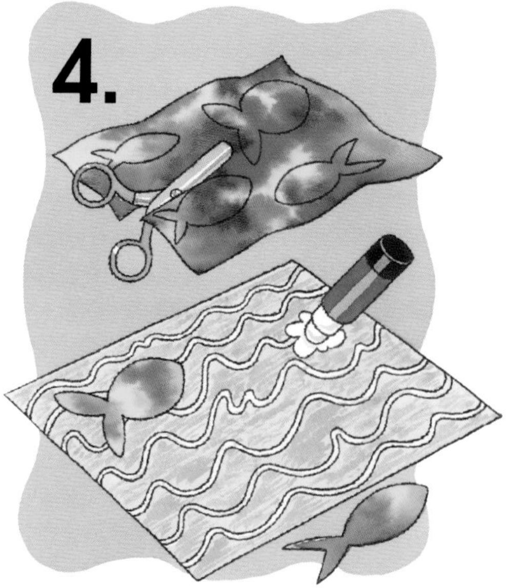

Cut out lots of fish from the painted paper towels. Glue them to the blue piece of paper.

*Draw some green seaweed with a crayon.*

*Add some eyes to your fish.*

# Hanging cones

## You will need:

colored construction paper

paints

large plate

adhesive tape

thick thread or wool

glitter glue

**1.**

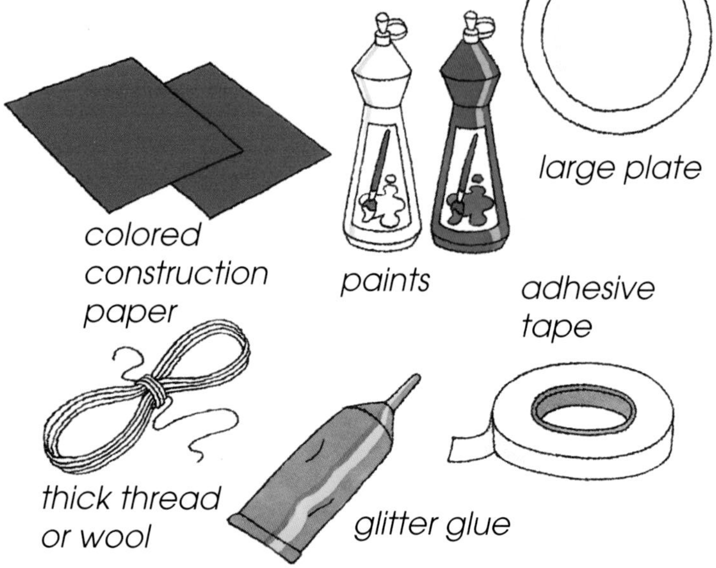

Draw around the plate on paper. Cut it out carefully.

**2.**

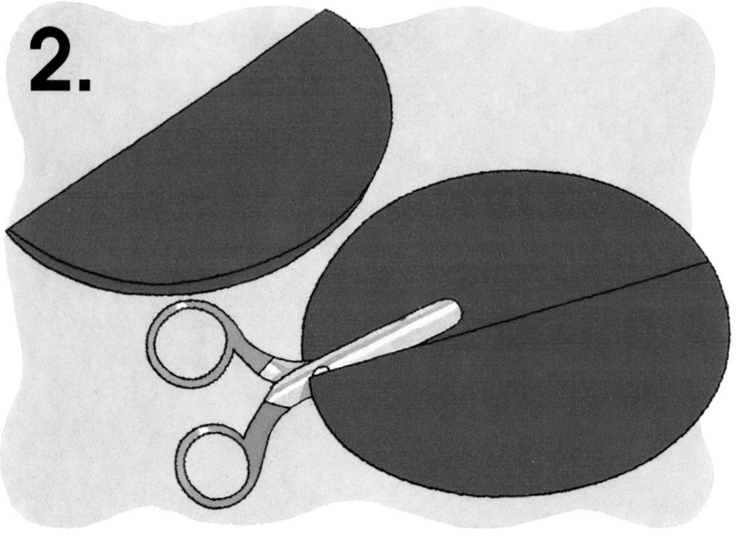

Fold the circle in half. Open it out and cut along the fold.

**3.**

Decorate one half circle with paint and glitter glue. Leave it to dry.

## 4.

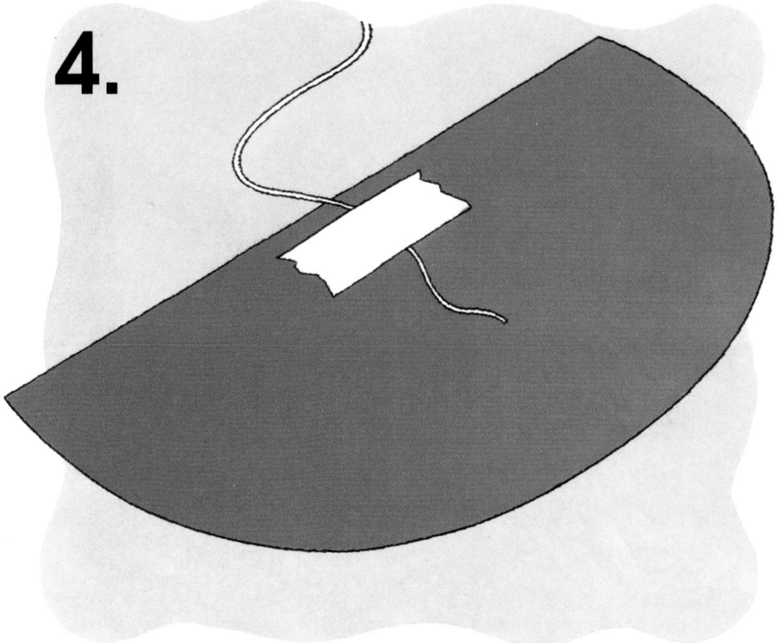

Now cut some thread and tape it to the back of the half circle.

## 5.

Overlap the edges to make a cone and tape down the edge.

*These make good decorations.*

*Make lots of cones.*

# Flick faces

## You will need:

large piece of paper

reusable putty adhesive

paints

round objects (mugs, cups, yogurt cartons)

scrap paper

**1.**

Draw around the round things on scrap paper. Cut the circles out.

*You can make your faces happy or sad.*

*Paint a pair of glasses on one of your faces.*

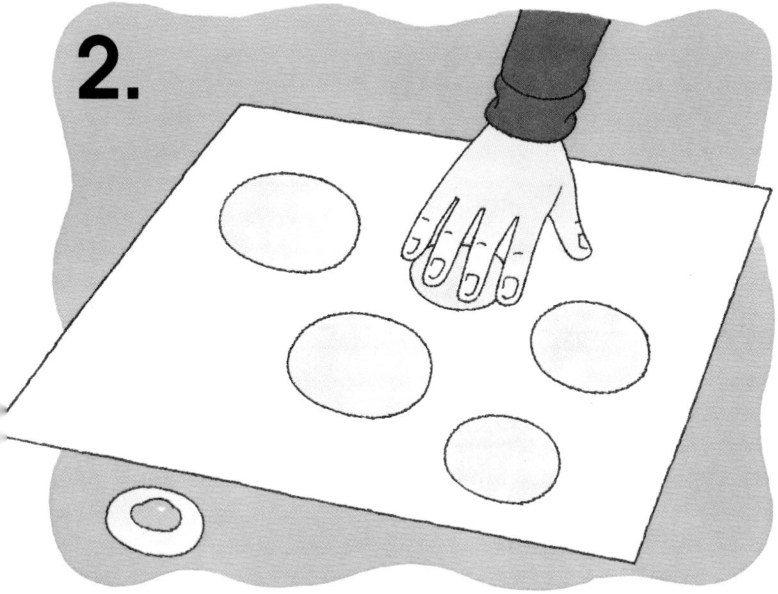


Stick the circles onto your paper with putty. Spread them out so that they are quite far apart.

Dip a thick brush in some paint. Flick the brush from the edge of a circle outward.

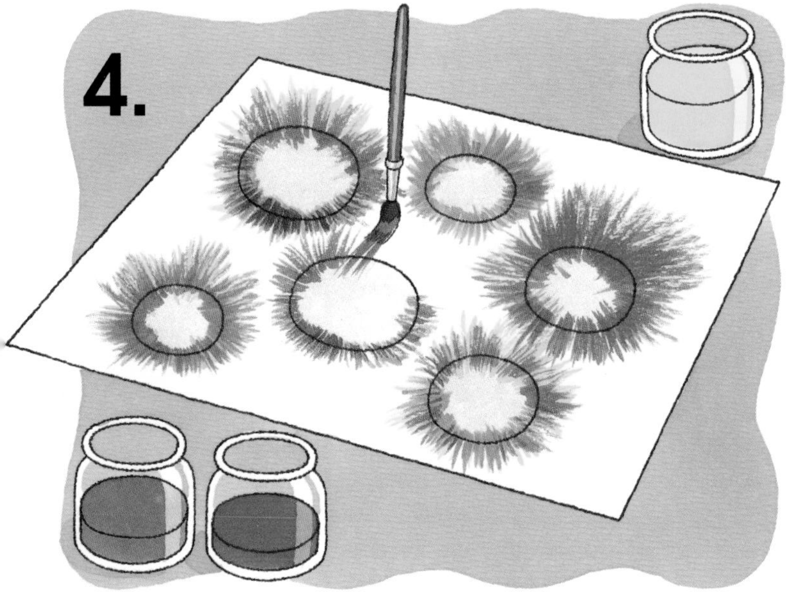


Carry on flicking paint around all the circles in this way. Add lots of different colors.

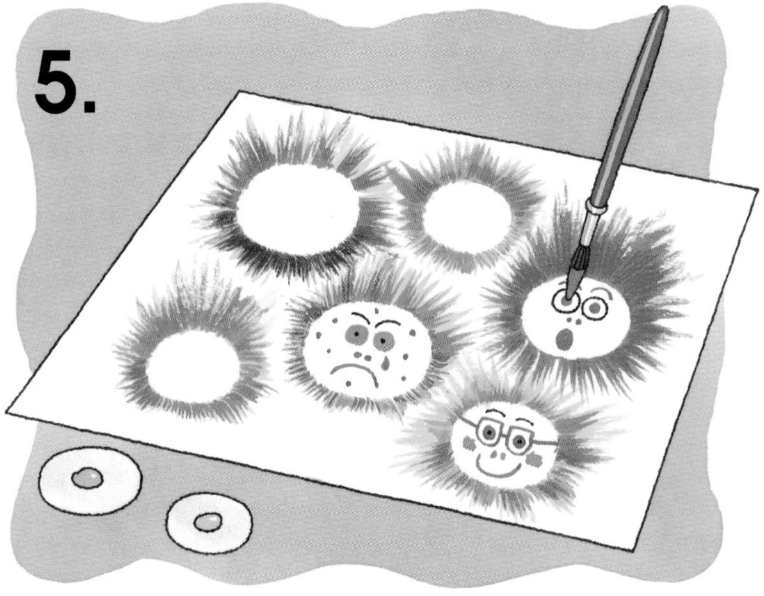


Peel off the paper circles and use a thin brush to paint in the faces. Leave to dry.

# Fierce lion

## You will need:

*large piece of paper*

*large plate*

*paint*

*mug*

Put the plate in the middle of the paper and draw around it.

Draw around the mug for eyes. Add ears, a nose, and a mouth.

Paint the lion's face. Add plenty of spots around its nose.

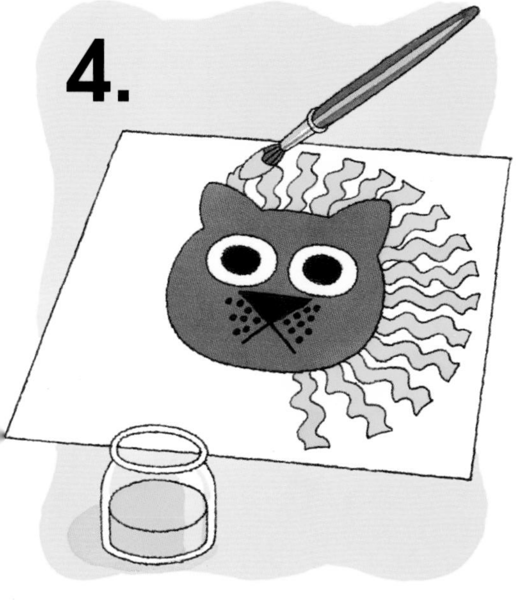


Paint wiggly yellow lines around the face for a mane.

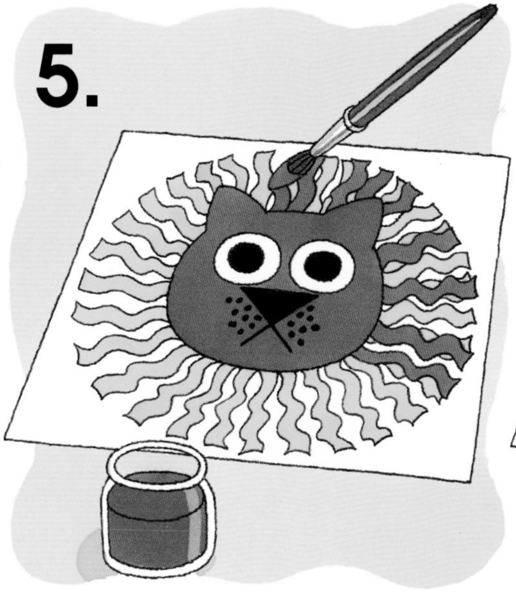


Add some light orange lines, then some darker orange in between.

Paint a line around the lion's face. Add lots of white whiskers.

# Frame it!

## You will need:

- *large paper, about a hand-width bigger than your painting*
- *bright paints*
- *scrap paper*
- *glue stick*
- *reusable putty adhesive*
- *small sponge*

*Try making other frames using different patterns.*

**1.**

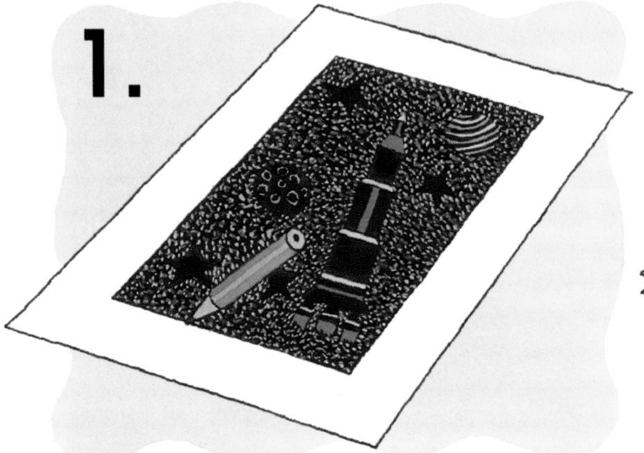

Draw around your painting on the large paper.

**2.**

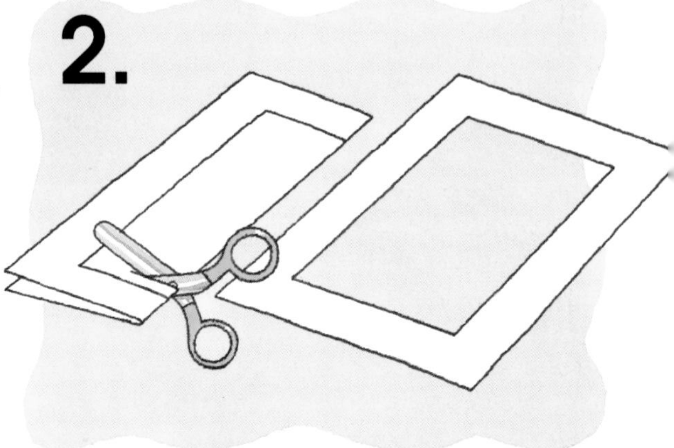

Fold the paper in half and cut just inside the line.

**3.**

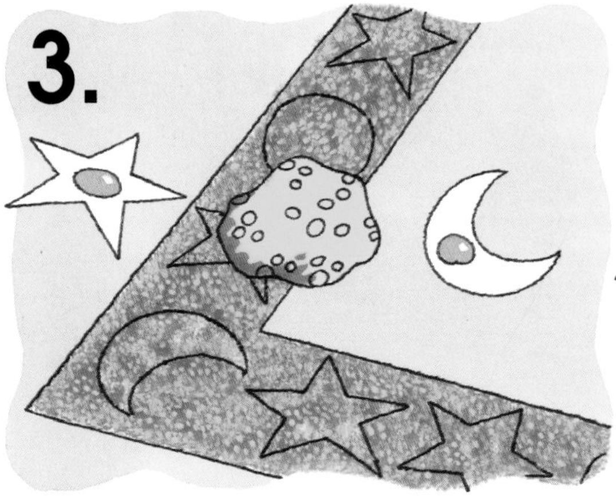

Stick on paper shapes with putty. Sponge on paints.

**4.**

Peel off the shapes and glue on the frame.

# Index